TROPICAL FRUITS

PERIPLUS NATURE GUIDES

TROPICAL FRUITS

Text and recipes by Wendy Hutton

Photography by Alberto Cassio

PERIPLUS

EDITIONS

lus Editions (HK) Ltd.

epublic of Singapore.
ISBN 962-593-135-X

Publisher: Eric M. Oey
Design: Peter Ivey
Text and recipes: Wendy Hutton
Photography: Alberto Cassio
with additional photography by Heinz von Holzen
(page 15 *left centre*, 25, 30 *top*, 39,
45 *top* and *left bottom*, 47 *top*, 49, 50 *top* and *left bottom*, 51 *left*)
Editor: Kim Inglis
Production: Mary Chia

Distributors

Indonesia
C.V. Java Books,
Jalan Kelapa Gading Kirana,
Blok A14 No. 17,
Jakarta 14240

Japan
Tuttle Shokai Inc.,
21-13, Seki 1-Chome, Tama-ku,
Kawasaki, Kanagawa 214

Singapore and Malaysia
Berkeley Books Pte. Ltd.,
5 Little Road #08-01, Singapore 536983

The Netherlands
Nilsson & Lamm B.V.,
Postbus 195, 1380 AD Weesp

United Kingdom
GeoCenter U.K. Ltd.,
The Viables Center, Harrow Way,
Basingstoke, Hampshire RG22 4BJ

United States
Charles E. Tuttle Co., Inc.,
RRI Box 231-5, North Clarendon,
VT 05759-9700

Introduction

One of the greatest joys of tropical Asia is the region's stunning array of ambrosial fruits. Their succulent sweetness, astonishingly varied shapes, colours, sizes and heady perfumes delight not only visitors, but also send the local residents into raptures as each fruit comes into season.

Pass a durian stall, where this notoriously pungent, spiky "king of fruits" is on sale—anywhere from Bangkok to Penang, Singapore to Bali, Zamboanga to Ambon—and you'll see enthusiastic crowds devouring the creamy rich flesh at makeshift tables or even just standing or squatting on the pavement, too eager to enjoy the fruit to wait to take it home.

Permanent local markets carry seasonal fruits as well as the many fruits are available year round, while in some areas, special fruit stalls are set up (especially near the orchards). It is common to find stalls at food centres and markets offering cut fruits, while itinerant vendors sell these as well as prepare special local fruit salads, often of tart or unripe fruits mixed with pungent sauces. While some countries such as Singapore have enviable hygiene standards, in other regions, travellers with sensitive constitutions are advised to avoid cut fruit.

As the countries of tropical Asia modernise at an increasing pace, some of the less commercial varieties of fruit—those that must be eaten within a day or so of ripening, others that are too fragile to be transported without considerable care—are difficult to find in major cities. Out in the smaller towns and villages, however, one can still come across unusual delights such as the fragrant "apple" of the cashew tree or the purplish-black Java plum.

This book is designed to act as a guide to the more common fruits found in tropical Asia. There are, of course, many others, some found only in cooler hilly regions, others available for a very brief season only.

Included at the back of the book are a selection of recipes illustrating some of the many ways in which fresh fruit may be served. Most people would agree, however, that nothing rivals a mouthful of meltingly smooth, juicy mango, or a wedge or two of sweet sour pineapple drenched with thirst-quenching juice. Indeed, it could be argued that to sample the fruits of tropical Asia is to experience a preview of Paradise.

Apple

MALUS SPP.

Botanical Family:
Rosaceae

Thai name:
Appoen

Malay name:
Buah apel

Indonesian name:
Buah apel

Tagalog name:
Mansanas

Apples are not readily associated with the tropics, yet the fruit is widely popular. Several tropical Asian countries have planted apples in cooler hilly areas in an attempt to meet local demand and reduce their dependence on imports from countries with temperate climates. Many would argue that the flavour does not match that of their Western counterparts, but it is possible nevertheless to find tasty crisp apples in Asia.

There are probably more apples grown in Indonesia than anywhere else in the region. Orchards planted in East Java, particularly in the hills around Malang, and a few in Bali produce green-skinned apples with red streaks. The flesh is almost pure white. There is also a very small variety, around 5 cm (2 inches) in diameter, sometimes referred to as a "cherry apple".

The apples are normally crisp and juicy, although they tend to be less sweet than temperate-climate varieties. They are quite pleasant when juiced.

Avocado

PERSEA AMERICANA

Avocadoes, native to Central and South America, are thought to have been introduced to Asia around two centuries ago and are particularly popular in the Philippines and Indonesia. Several varieties are grown, ranging in size from a pear-shaped fruit about 10 cm (4 inches) long to a round fruit weighing 1 kg (2 lbs). Unfortunately, Asian avocadoes rarely develop the rich, buttery flavour of the best varieties grown in Israel, Australia and California.

The green to greenish-black skin encloses pale yellowish-green flesh and a large beige stone. The creamy flesh is rich in oil, vitamins and protein. To eat, cut the fruit in half lengthwise and gently twist the two halves apart.

Asians often sprinkle sugar onto avocado halves and eat the flesh with a spoon, or put a little sweetened condensed milk into the hollow left by the stone. The flesh is also puréed to make a sort of milk shake, sweetened with evaporated or condensed milk. Avocadoes can be made into ice-cream and mixed with fruits to make fruit salad.

Botanical Family:
Lauraceae

Thai name:
Awokhado

Malay name:
Buah apukado, buah mantega

Indonesian name:
Buah apokat, avocad

Tagalog name:
Abukado

Botanical Family:
Musaceae

Thai name:
Kluai

Malay name:
Pisang

Indonesian name:
Pisang

Tagalog name:
Saging

Banana

MUSA SPP.

Those accustomed to the commercially popular bananas exported to the West—large pale yellow bananas with white, somewhat bland flesh and virtually no fragrance—are astounded by the range of sizes, shapes, skin colours and flavours of bananas found in tropical Asia.

The banana plant is native to the region and may be seen growing wild on hillsides and in secondary forests. The fruit of these is generally full of hard seeds and inedible. Bananas are high in food value, containing vitamins B and C, as well as minerals. Not only the fruit is utilised. Banana leaves serve as an all-purpose wrapping for steamed or baked food and as a disposable plate. The centre of the stem is treated as a vegetable and is the basis of what might be regarded as the Burmese national dish, a soup called *mohinga*. And after cooking, the centre of the banana bud tastes surprisingly like an artichoke.

Bananas range in size from tiny "lady's finger" bananas around 8 cm (3 inches) in length up to the giant "king" or "*rajah*" bananas as much as 45 cm (18 inches) long. The flavour varies from mild and sweet to slightly acidic, the texture from dense and slightly moist to floury, and the fragrance from faint to highly perfumed.

There are two categories of bananas: "dessert" bananas, which are sweet and generally eaten raw, and varieties that are always cooked. In some countries, this latter type is called plantain *(M. x paradisiaca)*. Dessert bananas tend to be green skinned when unripe, taking on a yellow colour when ready to eat. When over-ripe, the skin starts to develop brownish spots and at this stage, they are best used in banana milk shakes, or cooked in bread or cakes.

Cooking bananas are often treated as a vegetable, made into curries or sliced and deep fried. They are also coated with batter and deep fried, a popular snack throughout Southeast Asia; in Thailand the batter is likely to be made of rice flour and coconut milk. In the Philippines they are sprinkled with sugar and shallow fried in oil.

Breadfruit

ARTOCARPUS ALTILIS

Botanical Family:
Moraceae

Thai name:
Sa-ke

Malay name:
Sukun

Indonesian name:
Sukun

Tagalog name:
Rimas

The breadfruit, native to the Pacific and tropical Asia, grows on a large beautiful tree with dark green glossy multi-lobed leaves. The fruit, which averages about 2 kg (4 lbs) in weight, is generally spherical with green smooth skin which turns pale yellow when ripe. Easily grown, it is regarded as a rather ordinary, everyday plant.

In Asia, breadfruit is usually treated as a vegetable and therefore picked before it is ripe, when the skin is still green but with a slight yellowish tinge. The coarse skin is cut off and the interior cut in chunks and often par-boiled before being added to coconut milk and seasoning to make a curry-like dish. Breadfruit is sometimes deep fried in very thin crisp slices and sprinkled with either salt, chilli powder or sugar syrup.

There are two species of *A. altilis*: the seedless breadfruit and seeded breadnut. The seeds of the latter are excellent when boiled in salted water and may be served as a snack or a vegetable.

Camias

AVERRHOA BILIMBI

A member of the same family as the starfruit *(see page 48)*, this fruit is native to Southeast Asia and grows on a medium-sized tree in kitchen gardens in many parts of tropical Asia. Looking a little like a miniature cucumber but faintly five-angled in cross-section, the camias (also called the belimbi) has a thin smooth skin and is full of very juicy pulp. They are best bought when firm.

Because the flavour is extremely sour, the camias is rarely eaten raw but is used to make tangy sambals, pickles and chutney. It is also added to some curries to provide acidity (it is particularly good with fish) and can be used to make drinks and jam, provided plenty of sugar is used.

The fruit is generally very inexpensive and is often sold in plastic bags in Asian vegetable markets. In the past, the camias was used—as was the sour tamarind fruit—to remove rust and stains from knife blades, hands and clothing. Today, because sugar is cheaper and more readily available in Asia, the fruits are also eaten.

Botanical Family:
Oxalidaceae

Thai name:
Ta-ling pling

Malay name:
Blimbing asam

Indonesian name:
Belimbing wuluh

Tagalog name:
Kamias

Coconut

COCOS *NUCIFERA*

Botanical Family:
Palmae

Thai name:
Ma-phrao on

Malay name:
Kelapa

Indonesian name:
Kelapa

Tagalog name:
Niyog

The coconut, like the banana, is used in countless ways. The interior pulp of young coconuts is eaten raw, while the mature nut is grated and used in cakes and desserts. Or it may be squeezed with water to make coconut milk, the basis of so many regional dishes which, sadly, is high in cholesterol. Cooking oil is made from the mature flesh, while the flowers of the palm obligingly produce a sap which is made into an alcoholic drink; if this is left to ferment for several months, the result is coconut vinegar, widely used in the Philippines. Coconut sap is also boiled and made into a golden brown palm sugar.

The very centre or heart of the coconut palm, at the top near where the branches grow, is an excellent vegetable. It is common in markets throughout the Philippines, where it is known as *ubod*.

The unripe coconut, with either a green or yellow exterior depending on the variety, is full of faintly sweet water which makes a refreshing drink. This coconut water is sterile and has food value, so makes the perfect drink for invalids. An added bonus is the jelly-like flesh, which can be scooped out with a spoon. Rows of coconuts, their tops trimmed to make the opening easier, are a common sight throughout Asia. The vendor will slash the fruit open when it is purchased, sometimes adding a few ice cubes if such a luxury is at hand, for drinking on the spot.

The mature fruit is generally sold still with its hard, fibrous covering. Westerners, accustomed to seeing the coconut with this covering removed, leaving only the bright brown interior shell of the coconut on view, may fail to recognise ripe coconuts. The interior flesh of the ripe coconut is hard and, in Asia, is not eaten in chunks as a fruit but is grated and used either for coconut milk or as an addition to a variety of sweet and savoury dishes. The liquid inside mature coconuts is sometimes used to ferment various doughs for cooking, but most Asians believe it is bad for the health to drink it fresh.

Cashew Apple

ANACARDIUM OCCIDENTALE

Botanical Family:
Anacardiaceae

Thai name:
Mamuang-him-ma-phan

Malay name:
Jambu monyet, jambu golok, gajus

Indonesian name:
Jambu monyet, jambu mede

Tagalog name:
Kasoy, balubad

Everyone knows the excellent flavour of the cashew nut, but how many have tasted the juicy fruit of this plant, brought to Asia from tropical America by the Portuguese?

In fact, the true fruit is what is known everywhere as the nut, and the "fruit" sold for eating is a swollen stem.

The cashew tree is widely known in tropical Asia for its medicinal properties. All parts of the tree contain a sap which is irritant, including the thin membrane between the actual nut and its hard casing.

The cashew apple has a very thin skin—green when unripe and turning to yellow, pink, or more rarely, bright scarlet, when ripe. Because of its fragility, it is not a widely available "commercial" fruit. It can, however, be found at fruit stalls near cashew-growing regions. The slightly elongated fruit is about 7 cm (2 1/2 inches) long, with an interior of white flesh. Eat the fruit only when fully ripe or it is unpleasantly astringent. The ripe fruit is sweet, crisp and juicy with a faint rose perfume.

Custard Apple

ANNONA SQUAMOSA

Around nine varieties of this fruit, native to tropical America and introduced to tropical Asia several centuries ago, are cultivated around the world. Close relatives include the cherimoya *(A. cherimoya)* and sugar apple *(A. reticulata)*. The most commonly found Asian variety is quite small (about 8 cm or 3 inches in diameter). All fruits share the same distinctive appearance, with the skin composed of overlapping fleshy green "petals".

The interior has a very white sweet flesh, delicately flavoured with a hint of acidity, like its larger cousin, the soursop *(A. muricata)* *(see page 47)*. The custard apple is full of small segments of flesh containing shiny black seeds.

It is important to eat the fruit at exactly the correct stage of ripeness. It should still be slightly firm, yielding to gentle pressure with the palms. Avoid any fruits which feel soft as they will be over-ripe and therefore somewhat floury in texture. Because of its juiciness, the custard apple is ideal for drinks or desserts such as sorbets.

Botanical Family:
Annonaceae

Thai name:
Noi-na

Malay name:
Nona sri kaya

Indonesian name:
Srikaya

Tagalog name:
Atis

Dragon's Eye Fruit

DIMOCARPUS LONGAN

Botanical Family:
Sapindaceae

Thai name:
Lam-yai

Malay name:
Mata kucing, longan

Indonesian name:
Longan

The dragon's eye fruit is the true longan, one of many Southeast Asian natives which now grows in China, Taiwan and Thailand. A smaller fruit with very thin flesh, also called cat's eyes (*D. longan* spp. *malesianus*) grows only in Southeast Asia, and is a vastly inferior fruit.

Dragon's eye fruits are small and round and have a single shiny brown round seed inside. They grow in clusters and should be bought when still firm to the touch. The flesh is white, with a delicate flavour and a good balance of sweetness and acidity. There is a faint, indescribable fragrance to this fruit—almost woody but pleasant. To open, simply squeeze gently with the fingers.

Dragon's eye fruits are reported to maintain their flavour after cooking, though most people eat them raw. Dried, they have an intriguing smokey flavour and are sometimes used to make a fruit "tea". Tinned fruits are available, but their flavour—like that of tinned lychees—is a poor substitute for the real thing.

Duku

LANSIUM DOMESTICUM

Botanical Family:
Meliaceae

Thai name:
Long gong

Malay name:
Duku

Indonesian name:
Duku

The duku, native to Malaysia, is regarded there as one of the most delicious of all fruits—after, of course, the durian. Now cultivated throughout Asia, the duku grows on a tree belonging to the same family as the langsat *(see page 26)* and shares many similarities with that fruit.

The duku, which grows in clusters, is about the size and shape of a large golf ball, and has a thick, somewhat leathery golden brown skin. After gentle pressure with the hands on both sides of the top of the fruit (where the stem was attached), the skin splits to reveal five segments of white flesh, some of which may contain small seeds. Avoid biting into these as they are bitter.

The flesh is sweet and juicy, with perhaps a hint of grapefruit in the flavour, and aficionados claim it is superior to the langsat. Certainly, the lack of sticky latex in the skin typical of langsat makes eating a duku a much easier task. In Asia, the duku is always eaten fresh as a snack or to finish a meal.

Durian

DURIO ZIBETHINUS

Botanical Family:
Bombacaceae

Thai name:
Thurian

Malay name:
Durian

Indonesian name:
Durian

Tagalog name:
Durian

The durian, Southeast Asia's most highly prized fruit, is also its most controversial because of the overpowering odour. It is the only fruit banned from airline cabins, hotels and some public transport.

Native to Southeast Asia, the fruit of the very tall durian tree is roughly the size and shape of a spiky football. Inside the tough skin are five white segments enclosing two or three portions of soft cream-coloured flesh, each wrapped around a single large beige seed. Both the flesh and the seed (after boiling) are edible.

The durian is surrounded by folklore. It is reputed to be an aphrodisiac (there's a Malay saying that when the durians are down, the sarongs are up). It is also claimed to be dangerous to drink alcohol when consuming durians. The Chinese believe the durian is very "heaty" to the body (but this doesn't stop them consuming vast numbers of them just like everyone else!).

The famous 19th-century naturalist, Alfred Russel Wallace, described the durian thus: "It is like a buttery custard flavoured with almonds, intermingled with wafts of flavour that call to mind cream-cheese, onion sauce, brown sherry, and other incongruities.... It is neither acid, nor sweet, nor juicy, yet one feels the want of none of these qualities for it is perfect as it is."

Durians should be eaten within hours of their falling or being harvested, and fruits which have split open should be avoided as the flesh deteriorates quickly when exposed to the air. Roadside stalls spring up near durian orchards or in special markets in towns during the season. Most durian lovers cannot wait to take the fruit home and ask the vendor to open it (usually with the aid of a pair of gloves and a strong knife) so they can devour it immediately.

Today's durians are almost all hybrids and each has its special characteristics. Durian is best consumed fresh, although inferior quality or over-ripe fruit is also cooked to make sweetmeats such as *dodol* or made into jam.

Guava

PSIDIUM GUAJAVA

Botanical Family:
Myrtaceae

Thai name:
Farang

Malay name:
Jambu batu

Indonesian name:
Jambu biji, jambu kluthuk

Tagalog name:
Bayabas

The guava, a native of tropical and sub-tropical America, was known to the Aztecs as the sand plum, a reasonably good description of its somewhat gritty-textured flesh. Spanish *conquistadors* brought the plant to Asia centuries ago, along with the papaya, and it now flourishes around the world wherever the climate is suitable.

Several different varieties of guava are found in tropical Asia. All have a thin edible skin, with many small edible seeds embedded in the centre of the flesh inside. Many kitchen gardens have a guava tree as much for its medicinal properties as for its fruit, because the leaves—and sometimes the roots and bark—are universally known as a cure for dysentery. In the Philippines, water in which the fruit has been soaked is used to treat diabetes.

The most common guava in the region, and the only one grown on a large commercial scale (especially in Thailand and the Philippines), is also the least flavourful. This is the large green apple-sized guava. One reason for its popularity may be that you get more flesh in relation to seeds than in the smaller varieties. Apple guavas are crisp but not very sweet. Asians improve the bland flavour by dipping slices of the fruit into salt or chillies with soy sauce, while many Chinese like to sprinkle it with powdered dried plum. It is also good for pickles.

There are two types of golf-ball sized guavas, which are green skinned when unripe and take on a golden yellow hue when ready to eat. The more common of these has creamy-coloured flesh which is sweet and generally juicy, while the other has salmon pink flesh and a heady perfume.

Because of their high pectin content which helps the juice to set quickly, guavas are favoured for jams, jellies and preserves, particularly in the Philippines. Guava juice is also popular. The large apple guava can replace normal apples in various cooked desserts, but it seems a pity to diminish the delicate fragrance and sweetness of smaller guavas by cooking them.

Jackfruit

ARTOCARPUS HETEROPHYLLUS

Botanical Family:
Moraceae

Thai name:
Kha-nun

Malay name:
Nangka

Indonesian name:
Nangka

Tagalog name:
Langka

The jackfruit is said to be the largest fruit in the world, with the biggest specimens weighing up to 50 kg (110 lbs). Native to south India, the jackfruit is now firmly established all over Southeast Asia. The young fruit is cooked as a vegetable, often in coconut milk, while the ripe fruit provides almost enough for an entire village. The seeds make an excellent snack when boiled in salted water. Various parts of the plant are believed to have medicinal properties and the lemon-coloured wood is extensively used for wood carving, especially in Bali.

Ripe jackfruit, which has a rather strong odour, is generally cut into sections for sale in the market. The interior white central core is surrounded by yellow egg-shaped pieces of pulp, each of which contains a light brown seed. Many fruit sellers separate the pieces of pulp an them individually or by the bag. The flesh has a slightly chewy but agreeable texture, and a flavour that calls to mind pineapple mixed with melon.

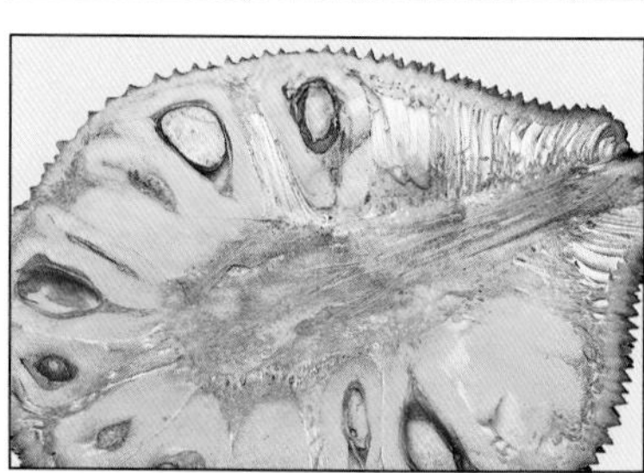

Java Plum

SYZYGIUM CUMINI

Baskets of purplish-black shiny fruit resembling black olives sometimes surprise the visitor to the Philippines, Thailand and parts of Indonesia. Their brief season limits opportunities of sampling the Java plum, and as the fruit crushes easily, it is never transported any great distance.

Inside the skin is translucent white flesh, making the fruit very much like a grape in texture. Like so many tropical fruits, the Java plum has medicinal properties, various parts of the plant being used to treat stomach aches, diabetes and ulcers.

It is important to eat the Java plum at the correct stage of ripeness. When not quite ripe, it is mouth-puckeringly sour because of the high tannin content. Some Filipinos, however, enjoy it at this stage, dipping it first in salt for a [illegible]y sour treat. Most would agree, however, that the fruit is better eaten ripe when it is juicy and sweet. Take care of the long seed in the centre: it stains everything it comes into contact with a purplish-red colour.

Botanical Family:
Myrtaceae

Thai name:
Look hwa

Malay name:
Jamun, jambolan

Indonesian name:
Jamblang, duwet

Tagalog name:
Duhat

Langsat

LANSIUM DOMESTICUM

Botanical Family:
Meliaceae

Thai name:
Langsat

Malay name:
Langsat

Indonesian name:
Langsat

Tagalog name:
Lansones

The langsat, like the closely related duku *(see page 19)*, is a native of western Southeast Asia, from Thailand to Borneo. It grows in clusters like bunches of large buff-coloured grapes. When the fruits are past their prime, they fall off the stalk and the skin takes on a brownish tinge. The thin skin can be peeled by hand but it has a sticky sap which is slightly unpleasant, although not irritant.

Inside the langsat are five segments of very juicy white flesh, the flavour being a delightful combination of acid and sweet. Some segments of flesh contain very bitter seeds, although through selective breeding, growers have managed to raise trees which produce langsat with very small seeds or even none at all. This type can generally be recognised by the even-shaped fruit which is slightly pointed rather than round or uneven in appearance.

The bark of the langsat tree is used to treat dysentery and scorpion stings. The seeds help lower fever and are also reputed to get rid of intestinal worms.

Lychee

LITCHI SINENSIS; NEPHELIUM LITCHI

The lychee, or litchi, native to southern China, is highly esteemed by Chinese people. The exiled Chinese poet, Su Tung Po, said the lychee was his only consolation when he was banished to the south.

The small fruits of the lychee are covered with a bumpy bright red skin; if you see any with the skin turning brownish, avoid them as they are over-ripe. The skin is easily stripped off to reveal a translucent white fruit in which is embedded a shiny, brownish-black seed. The flavour of the lychee is sweet, with just a hint of acid. The fruits are almost always eaten fresh, although Thai cooks also stuff them with savoury fillings. Everyone agrees that it is one of the nicest of fruits and the price reflects this—partly because many lychees sold in tropical Asian markets are flown in from China. The lychee is rather choosy about where it flourishes, and although it grows successfully in parts of Indonesia (especially in central Bali) and in parts of Thailand, it does not like heat and humidity.

Botanical Family:
Sapindaceae

Thai name:
Lin-chi

Malay name:
Laici

Indonesian name:
Lici

Tagalog name:
Licheas

Lime

CITRUS SPP.

Botanical Family:
Rutaceae

Thai name:
Ma-nao: large lime; Ma-krut: kaffir lime; Ma-nao-wan: kalamansi lime

Malay name:
Limau nipis: large lime; Limau purut: kaffir lime; Limau kesturi: kalamansi lime

Indonesian name:
Jeruk nipis: large lime, Jeruk purut: kaffir lime

Tagalog name:
Dayap: kaffir lime; Kalamansi: kalamansi lime

Three types of lime are found in most parts of Southeast Asia. The large lime (C. *aurantifolia*) is generally less elongated in shape than a lemon and has thin green skin which turns pale yellow when ripe. This thin skin is acknowledged in the Malay and Indonesian names for the fruit, which translate as "thin-skinned citrus".

The large lime is used mainly for its juice, made into a refreshing drink and sometimes added to cooked dishes for sourness and flavour. If ordering a fresh lime juice in Thailand, be aware that the Thais like both salt and sugar added, a surprising combination if you're expecting a sweet drink. The high vitamin C content of lime juice is widely known and it is considered good for general health as well as for the prevention of colds.

The rough dark green skin of the kaffir lime (C. *hystrix*) gives it a rather unattractive appearance, and the flesh inside is bitter and yields almost no juice at all. The secret of this fruit is the intense fragrance of its rind, which is often grated and added to cooked dishes, and its distinctive double leaf, used to add flavour to many Indonesian, Thai and Malay dishes.

The Philippines is the original home of the kalamansi (C. *mitis* var. *microcarpa*) or musk lime. This excellent lime is increasingly known abroad (often as calamansi) as it grows easily in any area where citrus fruits flourish. The round, thin-skinned kalamansi has green skin which turns yellow when the fruit is very ripe. The very juicy fragrant flesh is golden yellow. Kalamansi juice is highly prized for its intriguingly different flavour which some describe as "musky" and slightly less sharp than normal lime juice.

Malaysians, Thais and Singaporeans serve this lime to enhance the flavour of chilli-hot condiments or sambals and often squeeze it over noodle dishes. It is also used as a drink with added sugar and, in some areas, sour dried Chinese plums are partnered with kalamansi juice for an even more delightful flavour.

Malay Gooseberry

PHYLANTHUS ACIDUS; CICA ACIDA

Botanical Family:
Euphorbiaceae

Thai name:
Ma-yom

Malay name:
Cermai

Indonesian name:
Cermai

Tagalog name:
Iba

These fleshy little fruits, the family of which is probably native to India and Madagascar, are found in most parts of tropical Asia. They are often known as the gooseberries of that particular country; for example, Sri Lankan gooseberry, Malay gooseberry and so on. They grow in grape-like clusters and have greenish-yellow smooth skins; measuring around 2 cm (3/4 inch) in diameter, they have a single hard stone inside.

As the scientific name implies, these fruits are very sour and are never eaten raw. Sometimes, it is difficult to prise the flesh away from the stone. Firm, crisp and juicy when ripe, the Malay gooseberry is usually used to make preserves and pickles, but can also be cooked with sugar to make jams and jellies.

When cooked with sugar, the fruit changes colour—as does the temperate fruit, quince—from yellow to red. The Malay gooseberry can also be cooked together with apples, but be sure to adjust the amount of sugar to taste.

Mangosteen

GARCINIA MANGOSTANA

Like the famous durian *(see page 20)*, the mangosteen is a Southeast Asian native and bears fruit at the same time. The thick woody shell of the purplish-black mangosteen encloses several segments of the most exquisite juicy white flesh, sweet yet slightly acid. It is considered a perfect balance to the rich "heatiness" of the durian.

Unfortunately for lovers of fine fruit, the mangosteen is notoriously difficult to grow. And when the mangosteen does eventually bear fruit, the difficulty is not over as it bruises easily when transported. Mangosteens can be opened by squeezing gently on either side; if the skin does not pull apart easily, the fruit may not be ripe so wait another day or so. Alternatively, the top third of the fruit can be sliced off, revealing six or seven snowy white segments nestling in the thick pinkish-brown interior.

Like many tropical fruit trees, the mangosteen has its uses in folk medicine. The bark and skin are used to treat diarrhoea, and in Indonesia it is used to control high fever.

Botanical Family:
Guttiferae

Thai name:
Mangkut

Malay name:
Manggis

Indonesian name:
Manggis

Tagalog name:
Manggis

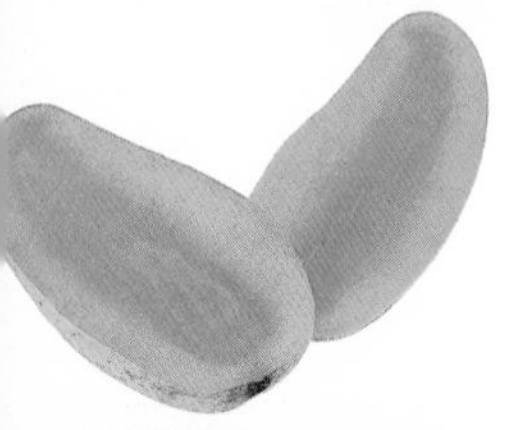

Mango

MANGIFERA SPP.

Botanical Family:
Anacardiaceae

Thai name:
Ma-muang

Malay name:
Mangga

Indonesian name:
Mangga

Tagalog name:
Mangang kalabau, mangga

There are dozens of varieties of mango family, varying in fragrance and flavour from sublime to unpleasant. Each country has developed its own varieties and a mango lover is hard-pressed to choose, say, between the very long, almost white-fleshed flower mango of Thailand, the small orange-fleshed Philippines mango dripping with sweet juice and the stronger-smelling and slightly sharp *arumanis* mango of Indonesia.

All these fruits are hybrids, as most of the varieties native to the region have somewhat stringy flesh with a sour, almost turpentine, flavour. These mangoes are generally made into a pickle or preserve.

The sap found in the leaves, stem and fruits of all types of mango is irritant and can cause a rash to those allergic to it; the Malaysian *kuini* is a particular culprit. Various parts of the mango—leaves, skin of the fruit, bark, seeds and resin—are used to treat many ailments, including diarrhoea and excessive bleeding.

Mangoes can be divided into two broad categories: those that are eaten green (unripe) and dessert mangoes enjoyed for their sweetness. Unripe mangoes are the perfect answer to the Asian love of sharp sour flavours. These fruit are peeled and eaten in salads or with savoury or chilli-hot dips; unripe mangoes are also cooked to make various pickles and chutney.

Dessert mangoes vary in size, skin colour and shape, some being fat, green-skinned and almost round, others being pale golden and slender, still others having a reddish tinge. All fruits have a large elongated seed inside and a non-edible skin.

Mango slices served with sweet sticky rice mixed with coconut cream are regarded as the ultimate dessert in Thailand and the Philippines. Their colour, texture and flavour combine well with milk products: mangoes make good ice-creams, yoghurts and soufflés, but as they are so good fresh from the tree this is like gilding the lily.

Melon

CUCUMIS MELO

Botanical Family:
Cucurbitaceae

Thai name:
Teng-lai

Malay name:
Buah semangka

Indonesian name:
Blewah, melon

Tagalog name:
Milon

It is believed that the melon originated in Africa. Several varieties of melon are grown in tropical Asia, the most popular being the honeydew melon. This has a pale green or ivory skin and juicy sweet flesh.

The species to which the honeydew belongs is readily cross-pollinated with other varieties, and a number of hybrids are found in different markets in the region. Some of these look similar to the musk melon, which is characterised by a skin covered with netting, while others are closer to the European cantaloupe.

Casaba melons imported from China (*hami* is the most favoured) and from Taiwan (the *shiang gwa* or fragrant melon) are found in many markets throughout Asia, particularly in the larger cities.

Although melons are generally eaten in slices, either as a snack or at the end of a meal, they are also used to make juices and combined with coconut milk and sometimes pearl sago to make a delightfully refreshing dessert.

Nipa Palm Fruit

NYPA FRUTICANS

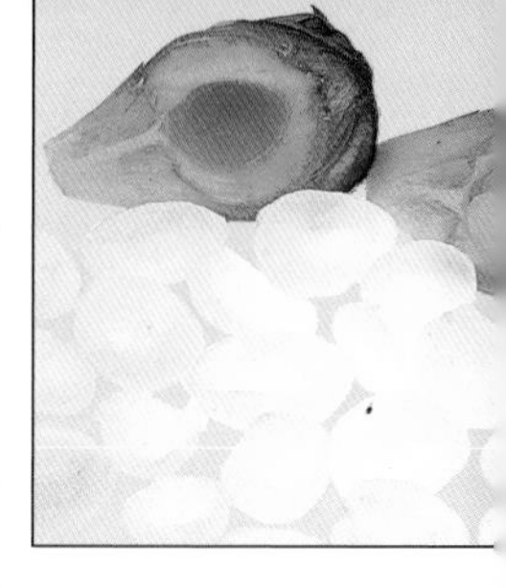

This beautiful palm, which grows in swampy areas and along rivers in Southeast Asia, is remarkably versatile, as indeed are many other palms such as the coconut and sugar palm. The dried fronds of this palm, known as *atap* in Malaysia and Indonesia, and *nipa* in the Philippines, are used to make a common roofing material, and can be transformed into baskets, mats and hats.

Before the opening of the nipa palm flowers, the infloresence can be tapped to obtain a sweet sap which is then boiled to produce a type of palm sugar. The same sap can also be left to ferment and produces an alcohol or toddy, and if this is left for several months, the result is a vinegar used in cooking.

The young shoots of the nipa palm are edible, as is the pulp of the immature seeds. These are white and slightly sweet, although are normally boiled in sugar syrup to make them more palatable. The nipa fruit can be eaten simply with its syrup, or mixed with other fruits.

Botanical Family:
Palmae

Thai name:
Chaak

Malaysian name:
Buah atap

Indonesian name:
Buah atap

Orange

CITRUS SPP.

Botanical Family:
Rutaceae

Thai name:
Som

Malay name:
Oren

Indonesian name:
Jeruk manis

Tagalog name:
Dalandan, kahel

There are over 100 varieties of the common sweet orange, each varying in size, colour and flavour. The orange is believed to have originated in China, hence one species name, C. *sinensis*. Although oranges imported from temperate countries are often found in tropical Asian markets, locally grown oranges can also be found.

These are usually green-skinned, with bright yellow to orange-coloured flesh. If the skin has turned from green to orange, this means the fruit is over-ripe and may be a little dry. Tropical oranges are normally juicy, but also contain a large number of seeds and can sometimes be somewhat sour. Some varieties are very easy to peel, although the firm Thai oranges can be an exception to this.

Mandarins (C. *reticulata* var. *nobilis*) and tangerines (C. *reticulata* Bl.) are popular for their sweet flesh and easy-to-peel skin; most of those found in tropical Asia have been imported from China where they symbolize good luck and are given as presents at the lunar New Year.

Otaheite Apple

SPONDIAS CYTHERA

This fruit is native to Southeast Asia, and is widely found in some Pacific Island countries (the name Otaheiti is the old name for Tahiti). It might be mistaken for a green-skinned mango, as it has the elongated shape of that fruit, as well as a similar resinous smell and smooth skin. It measures from 5 to 9 cm (approximately 3 ½ inches) and is sometimes called the ambarella.

Inside the Otaheiti apple there is a star-shaped central core with five small, pale green seeds; this is noticeable if the fruit is cut in horizontal slices.

The oval fruit has a relatively thick rind and the flesh inside is hard and crisp, with a tangy taste when still unripe. When ripe, however, the flavour can be quite good and tastes something like a mango-flavoured apple. Some varieties, however, are not sufficiently sweet to make good desserts, and for this reason the Otaheiti apple is often used in tangy sour salads, especially in Thailand and Indonesia.

Botanical Family:
Anacardiaceae

Thai name:
Ma-kok-farang

Malay name:
Kedongdong

Indonesian name:
Kedongdong

Tagalog name:
Hevi

Papaya

CARICA PAPAYA

Botanical Family:
Caricaceae

Thai name:
Ma-la-kaw

Malay name:
Betik

Indonesian name:
Papaya

Tagalog name:
Papaya

The papaya (known in some countries as the papaw), is renowned for is high vitamins A and C content and for its medicinal properties, while papain—an enzyme found in both the fruit and leaves—is used as a meat tenderiser.

A native of Mexico, there are many different varieties, ranging in size from long pendulous fruits of 35–40 cm (14–16 inches) to dainty little egg-shaped fruits. The skin is green, usually turning to yellow or orange as the fruit ripens. Inside the papaya is a cavity which contains a mass of shiny black seeds, which are discarded (although they are edible). The ripe flesh, ranging from golden to salmon pink, is always eaten raw, usually accompanied by a squeeze of lime juice; however, unripe papayas are used in salads or cooked as vegetables throughout Southeast Asia.

Many of the modern commercially grown hybrids, which are usually smaller in size than the old-fashioned fruit and known as Hawaiian or Solo papayas, have a richer flavour than older varieties.

Passionfruit

PASSIFLORA SPP.

Botanical Family:
Passifloraceae

Thai name:
Saowaros

Malay name:
Buah susu

Indonesian name:
Markisa, konyal

Tagalog name:
Pasionaria

Two varieties of this Brazilian native are cultivated in a few areas of tropical Asia. Because of its limited geographical range, it is not very common.

The Indonesian variety (*P. edulis forma edulis*) has two forms—one with a purple skin called *markisa* and one with a yellow-orange shell with a mass of juicy translucent pulp surrounding edible greyish seeds called *konyal.* The fresh raw fruit, with a sweet, almost milky flavour, can be scooped out with a spoon. The variety grown in the Philippines, Malaysia and Thailand (*P. edulis forma flavicarpa*) is egg-shaped with a lemon-coloured shell, golden flesh and shiny black seeds. If picked too soon, the flavour can be a little sharp and acidic.

The passionfruit was named by Spanish missionaries who thought that the striking purple flowers symbolised the passion of Christ, the three central stigmas representing the three nails, the feathery corona the crown of thorns, and the five anthers the five wounds.

Persimmon

DIOSPYROS KAKI

Botanical Family:
Ebenaceae

Thai name:
Plapchin

Malay name:
Buah kesemek, buah kaki

Indonesian name:
Kesemek

Native to China and very popular there and in Japan, these beautiful bright orange-skinned fruits are somewhat similar to a tomato in shape and size. The stem end of the fruit has four brownish, leathery "petals".

Persimmons are grown in some areas of tropical Asia and either the local or imported varieties can be found in markets and shops at the beginning of the northern hemisphere autumn season.

There are several varieties of persimmon—some hard, some softer—and all with bright orange flesh inside. The skin is difficult to digest and should not be eaten. The most popular way to serve a persimmon is to cut it in half and use a spoon to scoop out the smooth sweet flesh. The fruit must be properly ripe for it to have its full flavour, but over-ripe fruits should be avoided as the flesh will be mushy. Persimmons can also be added to fruit salads or blended to make juice; alternatively, try making a purée and add it to ice-creams or to make a fool or mousse.

Pineapple

ANANAS COMOSUS

The pineapple, native to South America, is cultivated throughout tropical Asia. The name comes from the Spanish word for pine cone (*piña*), which the fruit vaguely resembles with its scaly skin. Ripe pineapples have a juicy sweet flesh with just a hint of acidity to make them even more refreshing. Pineapple is not only good raw or cooked in savoury dishes, it also makes good pickles, chutney and jam, as well as delicious juice.

Several types of pineapple are found in the region. Some are grown only for ornamental use, their decorative leaves making them a popular pot plant. Small varieties that tend to be somewhat acid, or unripe fruits, are used as a vegetable or in sour fruit salads—and also made into pineapple curry. Freshly peeled and sliced ripe pineapple is found everywhere in the region. If buying a whole fruit, check that it is ripe by smelling to see if it is fragrant and try to tear one of the leaves sprouting from the top. If it comes away easily, the fruit is ready to eat.

Botanical Family:
Bromeliaceae

Thai name:
Sappa-rot

Malay name:
Nanas

Indonesian name:
Nanas

Tagalog name:
Piña

Pomegranate

PUNICA GRANATUM

Botanical Family:
Punicaceae

Thai name:
Thap-thim

Malay name:
Buah delima

Indonesian name:
Delima

Tagalog name:
Granada, delima

The pomegranate, which has been cultivated in Asia since ancient times, grows on a shrub or small tree, which is a favourite potted plant for many urban Chinese. The fruit is round and approximately the size of a small orange. Although the flowers are a striking orange-red, the outside of the pomegranate is an unexciting brownish-yellow with a blush of orange. Split open the skin, however, and you'll be rewarded with six segments of brilliant ruby-like seeds. The seeds or kernels are covered with a small amount of edible pulp. This is best made into juice by soaking one part of seeds with two parts of sugar for 24 hours, then bringing it to the boil.

As the high tannin content of the pomegranate makes it somewhat bitter and suitable mainly for juice, in Asia the fruit is grown more for its decorative and medicinal properties than for eating. The pulp is also made into the cordial, grenadine; used in cocktails worldwide, its name derives from the French word for the fruit, *grenade*.

Pomelo

CITRUS MAXIMA

The biggest fruit of the citrus family, the pomelo is thought to have originated in the bio-geographical area called Malesia (which stretches from Malaysia, through Indonesia to Papua New Guinea).

It is a large round fruit, about the size of a small football, and usually tapers to a slight point at one end. It has a very thick skin, light green turning to lemon yellow as the fruit ripens. Inside the fruit are large segments of pulp covered by tough membranes which need to be peeled off. Some pomelos are almost seedless, while others contain an annoying amount of seeds.

The flesh, which can be either pale lemon or a lovely strawberry colour, varies in sweetness and juiciness. The very best pomelos (and those produced in Thailand are uniformly good) are excellent; disappointing ones can be dry and somewhat tasteless. Asians get around the problem of less-than-perfect pomelos by treating them as a salad vegetable or dipping them in chilli-hot sauce.

Botanical Family:
Rutaceae

Thai name:
Som-o

Malay name:
Limau besar, limau Betawi, limau Serdadu

Indonesian name:
Jeruk Bali

Tagalog name:
Lukban

Rambutan

NEPHELIUM LAPPACEUM

Botanical Family:
Sapindaceae

Thai name:
Ngoh

Malay name:
Rambutan

Indonesian name:
Rambutan

Tagalog name:
Rambutan, usan

The name of this fruit comes from the Malay word for hair, *rambut*. It is a particularly apt description, for the rambutan looks like a bright red golf ball covered with whiskery hairs. These hairs are flexible and not sharp.

A member of the same family as the lychee and longan *(see pages 27 and 18)*, the rambutan looks somewhat similar inside: an oval of white flesh with an elongated smooth stone embedded in the centre. Some varieties of rambutan are almost freestone, but in others some of the flesh clings to the stone. You can ask to sample a fruit before committing yourself to buying. There is also a variety of rambutan with golden skin, although the interior of the fruit is no different from the red-skinned variety.

The flavour of a good ripe rambutan is sweet with a mere hint of acid, and it is somewhat juicier than the lychee. Rambutans are rich in vitamin C. To open the rambutan, twist it with both hands; luckily the absence of sap makes it pleasurable to eat straight from the hand.

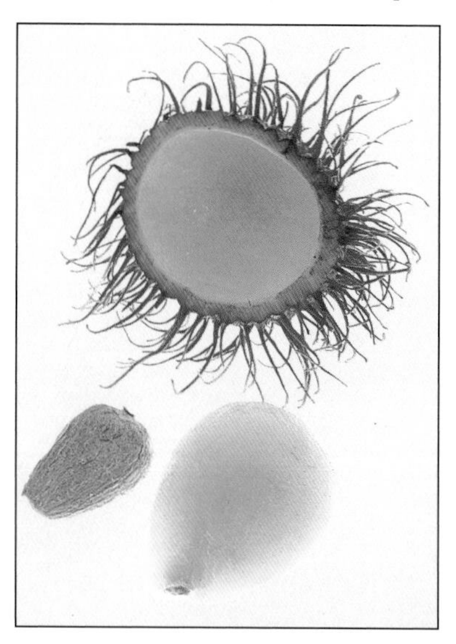

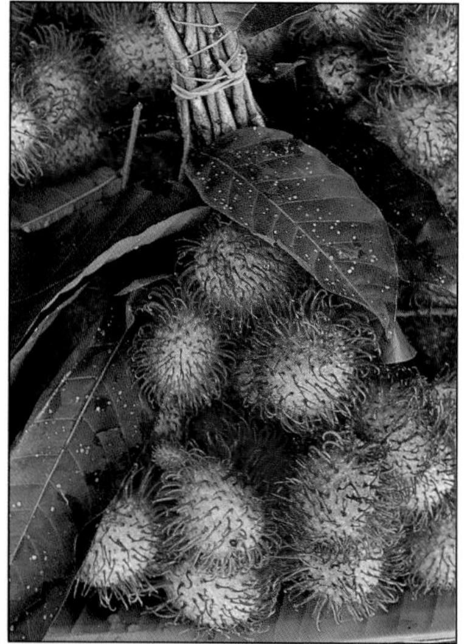

Salak

SALACCA SPP.

Botanical Family:
Palmae

Thai name:
Sala, rakum

Malay name:
Salak

Indonesian name:
Salak

The alternative English name for this fruit—snake fruit—is a good description of the appearance of its scaly brown skin. The salak grows on a short palm tree and is native to Indonesia. It also grows in parts of Malaysia and Thailand.

The tough skin of this golf ball-sized fruit is quite thin and strong and is easily peeled off. Inside are three or four segments of firm, creamy-beige flesh, each of which encloses a bright brown stone. Salak fruit has a high tannin content and if the fruit is not properly ripe, it can be unpleasantly astringent. The flavour of a ripe salak is intriguing, neither sweet nor sour, faintly nutty, with an excellent crisp texture. It has very little juice so can be eaten without fear of making a mess of one's hands or clothes. If the fruit is somewhat sour, it is best added to the sour fruit salads enjoyed by many Asians.

Aficionados of the salak claim that those grown near Sibetan on the slopes of Bali's Mount Agung are the best to be found anywhere in Asia.

Sapodilla

ACHRAS SAPOTA

Botanical Family:
Sapotaceae

Thai name:
Lamut

Malay name:
Ciku

Indonesian name:
Sawo manila

Tagalog name:
Chico

If you eat an unripe sapodilla, you'll find it tasteless and astringent, and should you be unlucky enough to eat an over-ripe fruit, you'll totally disbelieve the claim that it is one of the best native fruits of tropical America. But eat a perfect sapodilla and you'll think you are eating the finest pear flavoured with maple syrup which, as a French botanist said, smells of "honey, jasmine and lily of the valley".

The sapodilla (sometimes known as the naseberry) looks like a soft tan egg. Inside the skin, which should be peeled off with a knife, are segments of golden-brown flesh, each with a shiny black seed inside. The sapodilla, which has the faintly gritty texture of a pear, has too pronounced a flavour to mix with any other fruits and is best enjoyed raw, and on its own.

If the unpeeled fruit does not smell fragrant and is still hard, wait a day or two for it to ripen. It should still be slightly firm when pressed lightly but not be totally soft or the flavour will have deteriorated.

Soursop

ANNONA MURICATA

Although wedges of this tropical American native are sometimes served for eating with a spoon, most soursops in Southeast Asia are used to make juice or desserts such as mousse, ice-cream and jelly.

The slightly bumpy thin skin of this irregularly shaped fruit is green even when the fruit ripens. They are large, weighing in excess of 3 kg (7 lbs). Inside, the flesh is white and pulpy, full of shiny black seeds, with a central pithy core running its length. The soursop bruises easily when fully ripe, so buy it while still firm and wait until it yields slightly to gentle pressure. Then, eat it immediately.

The flavour of the soursop is somewhat acidic, but this is easily counteracted by adding sugar. It is refreshing, with a faint fragrance and an elusive but irresistible flavour. Soursop derives its name from the Dutch *zuur zak* or sour sack. Sop is an English word meaning something which soaks up liquid; as the flesh of the soursop is certainly saturated with juice, the name is not inappropriate.

Botanical Family:
Annonaceae

Thai name:
Thu-rian-khaek

Malay name:
Durian blanda

Indonesian name:
Sirsak

Tagalog name:
Guayabano

Starfruit

AVERRHOA CARAMBOLA

Botanical Family:
Oxalidaceae

Thai name:
Ma-fuang

Malay name:
Belimbing manis

Indonesian name:
Belimbing manis

Tagalog name:
Belimbing

The starfruit, which has an excellent crisp texture and is full of juice, is a native of Southeast Asia. Sometimes referred to as carambola, the name starfruit is apt, for if you cut a cross-section of this fruit, it looks just like a five-pointed star.

The uncut starfruit is a little like an elongated egg-shaped torpedo with five fins. The waxy skin is green in the unripe fruit, turning to a pale yellow or even rich gold as it ripens. The fruit should be eaten while still firm for it quickly loses its flavour once it softens. The edge of each ridge of the starfruit is bitter and should be sliced off before the fruit is served. The central pith, which may contain a few small seeds, can be removed during eating.

Most starfruit are somewhat tart in flavour, but a hybrid known as Honey Starfruit is sweet and fragrant. Starfruit is widely enjoyed as a dessert fruit and in Asian savoury salads. Rich in both vitamins C and A, it is reputedly a good cure for hangovers.

Sugar Palm Fruit

ARENGA PINNATA

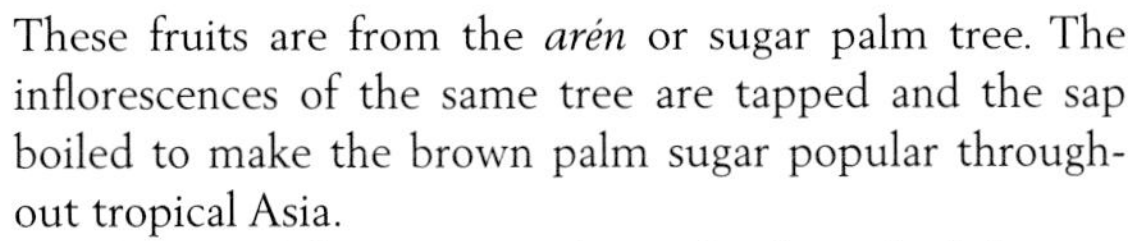

These fruits are from the *arén* or sugar palm tree. The inflorescences of the same tree are tapped and the sap boiled to make the brown palm sugar popular throughout tropical Asia.

The round fruits grow in long, closely packed clusters. A tough green skin covers a fibrous layer which contains three white pods. The fruits are always extracted from their pods and boiled before being sold. The juices of the raw fruit are highly irritant, although the fruits are perfectly safe after cooking.

Looking like elongated pieces of hard white jelly (and tasting much the same), cooked sugar palm fruits are kept in water until sold. They do not have much flavour on their own, so are often added to mixed ice concoctions (*es campur)* in Indonesia, Singapore and Malaysia and are used in a similar fashion in the Philippines.

Sugar palm fruits are often sold tinned or in jars, sometimes packed in sugar syrup.

Botanical Family:
Palmae

Thai name:
Lukchid

Malay name:
Kabung

Indonesian name:
Kolang kaling, buah enau

Tagalog name:
Kaong

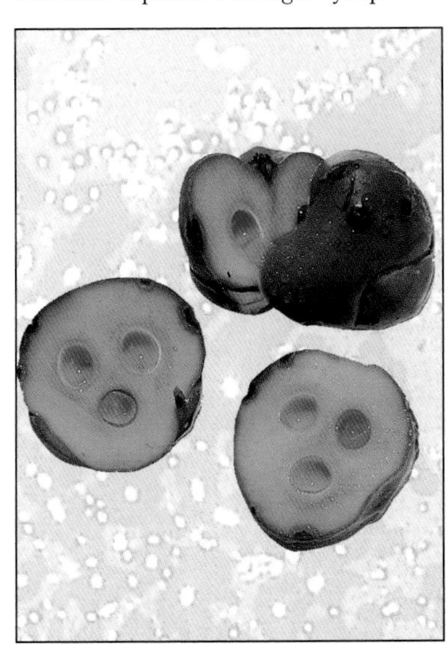

Water Apple

SYZYGIUM AQUEUM

Botanical Family:
Myrtaceae

Thai name:
Chomphu pa

Malay name:
Jambu air

Indonesian name:
Jambu air

Tagalog name:
Tambis

The water apple is a member of the myrtle family, native to Southeast Asia. It measures about 2.5–5 cm (1–2 inches) in diameter and grows easily on moderate-sized trees throughout the region. It is an inexpensive fruit.

Water apples have a very thin, almost waxy-looking edible skin, with crisp white flesh inside. In the centre is a rather spongy portion with a single seed. The skin colour ranges from palest green through a delicate blush pink to deep crimson and a sort of brownish red.

Water apples generally have only a faint flavour with a hint of sweetness and are prized mainly for their crisp juiciness. Pleasant to bite into raw, they also go well in all types of salads. A related species native to Malaysia, the Malay apple (*S. malaccense)*, is barrel-shaped and has green skin with blush-like patches of pink. It has a delightful fragrance while the flavour and texture are similar to the water apple. Another species is the rose apple (*S. jambos*), a more insipid fruit, used in preserves.

Watermelon

CITRULLUS LANATUS

This fruit, native to Africa, is now found throughout the tropical and sub-tropical countries of the world. There are several varieties of watermelon. They can be either round or elongated; some have dark green skin, others have pale green skin streaked with dark green.

Although the flesh is most commonly deep pinky red, a bright yellow variety (popular with the Chinese for its association with gold and wealth) has been developed in recent times. The golden-fleshed watermelon tastes the same as the pink variety, but generally costs more.

Some varieties of melon have been bred so as to be virtually seedless, although the toasted salted seeds are a snack throughout Asia, particularly among the Chinese.

Watermelons have the highest moisture content of any fruit (over 90%) and although they have little nutritional value and sometimes only a moderate amount of flavour and sweetness, they make a most refreshing snack during the heat of the day.

Botanical Family:
Cucurbitaceae

Thai name:
Taeng-mo

Malay name:
Tembikai

Indonesian name:
Semangka

Tagalog name:
Pakwan

Fruit Fool

This traditional English recipe for combining puréed summer fruits with sugar and whipped cream can be used successfully with many tropical fruits. Although cream gives the fool a richer flavour, evaporated milk can be used if cream is not available. Suggested fruits, used either on their own or in combination with other fruits, include soursop, custard apple, mango, avocado, banana, persimmon, durian or papaya.

1 1/2 cups puréed fruit
about 1/3 cup sugar, or to taste
1 1/2 cups chilled cream or evaporated milk

Mix the fruit with sufficient sugar to sweeten it. Beat the chilled cream or evaporated milk until stiff, then fold it into the fruit. Chill before serving.

Fruit Mousse

This is a good way to use very ripe soft fruits, mixing them with beaten egg white, gelatine and cream for a light dessert. Soursop is perhaps the best fruit of all to use for this, although banana, custard apple, avocado, mango, persimmon, durian or papaya could also be made into a mousse. Consider also mixing a couple of fruits together.

1 1/2 cups fruit pulp
1/4 cup sugar, or more to taste
lime or lemon juice to taste
3 teaspoons gelatine
1/2 cup warm water
1 cup cream, beaten
2 egg whites, stiffly beaten

Stir the pulp and sugar together to dissolve and add lime or lemon juice to taste.

Sprinkle the gelatine over the warm water and allow to soften. Put into a small pan and stir over low heat until the gelatine is completely dissolved. Pour into the fruit pulp and mix well.

Fold in the beaten cream, then carefully fold in the egg whites. Chill before serving.

Fruit Shake

This can be made with plain milk, yoghurt and water or coconut milk, with fruits and flavourings added to taste. Try to include at least 1 ripe banana for its texture, then add about 1/2 cup of other fruit such as soursop, custard apple, avocado, persimmon, mango, durian or papaya. The quantities given are enough for 2–3 glasses.

3 cups plain or coconut milk (or 1 cup yoghurt and 2 cups iced water)
a few ice cubes
1 ripe banana
1/2 cup soft fruit
1–2 tablespoons sugar, or more to taste
optional: flavourings such as 1/2 teaspoon vanilla essence, 1 tablespoon rum, freshly grated nutmeg, cinnamon or cardamom powder

Combine all ingredients in a blender and process until you have a thick creamy shake.

Ice-cream

This is a simple recipe, less rich than the normal ice-cream which calls for egg yolks and cream. However, the richness of the avocado more than compensates for the lack of cream.

250 g (8 oz) sugar
2 1/2 cups milk
1 small vanilla pod, slit, or 1 teaspoon vanilla essence
1 teaspoon instant coffee powder or 1 tablespoon dark rum
4–5 medium sized ripe avocadoes, mashed

Heat the sugar, milk and vanilla pod or essence in a pan, stirring until the sugar is dissolved and the mixture comes to the boil. Remove from heat, add instant coffee or rum and set aside to cool.

Add the avocadoes and blend well. Pour into a tray and freeze for 1 hour. Mix in a food processor or blender to break up any crystals, then put back in the tray and freeze until set.

Jam

This recipe provides a way of using sour fruits such as camias or Malay gooseberry.

2–3 cups ripe camias or Malay gooseberry
equal volume of sugar
1 cinnamon stick (8 cm/3 inches)
2 whole cloves
1/4 teaspoon cardamom seeds

Prick the fruits all over with a fork and leave to soak in water for about 4 hours. This will help draw out some of the sourness.

Drain the fruits, squeeze lightly with the hands and chop coarsely. Put into a non-aluminium pan with an equal volume of sugar and the spices. Bring slowly to the boil, stirring to dissolve the sugar, then simmer uncovered until the mixture is thick. This should take about 15–20 minutes, and the colour of the jam will have taken on a reddish hue.

Bottle in sterilised jars and refrigerate the jam after opening in a tropical climate.

Mango Chutney

Fruit chutney, a legacy of the colonial British rule in India, is a great way to use certain fruits. Without doubt, the most popular is mango chutney.

750 g (1 1/2 lbs) unripe green mangoes (about 6–8)
1 tablespoon salt
6–10 dried chillies
30 g (1 oz) fresh ginger
4 cloves garlic
3 tablespoons oil
1 teaspoon black mustard seeds
1/2 teaspoon turmeric powder
3/4 cup malt vinegar
1 cup soft brown sugar
1 heaped tablespoon raisins

Peel the mangoes, cut into thick slices and mix with salt. Cut off the stalk end of the chillies, shake out most of the seeds and cut the chillies into small pieces. Soak in hot water until soft, then blend chillies, ginger and garlic together with a little of the oil.

Heat remaining oil in a non-aluminium pan and put in the mustard seeds. When they start to pop, add the ground mixture and sauté over a low heat for 3–4 minutes. Add vinegar, sugar and turmeric, bring to the boil and simmer uncovered for 10 minutes.

Add the mangoes and simmer for about 15 minutes until the mangoes turn transparent. Add the raisins and put the chutney into sterilised jars. When cold, cover with a firm-fitting lid and store.

Melon and Sago dessert

It's hard to think of a better fruit for this refreshing light dessert than honeydew melon, although other types of melon (except for watermelon) could be substituted. Mango or soursop would also do well, although the amount of sugar may need to be increased if you use the soursop.

3/4 cup pearl sago
7 cups water
1 cup sugar
1/2 cup water
1/2 honeydew melon, or 3 ripe mangoes or 1/2 soursop
1 cup coconut milk

Soak the sago with 2 cups of water for 30 minutes, then drain. Bring the remaining 5 cups of water to the boil and add sago. Cook, stirring occasionally, until the sago balls are transparent. Drain in a sieve and wash under cold running water. Leave aside to cool.

Boil together the sugar and 1/2 cup water to make a syrup. Leave to cool.

If using melon, peel, and discard the seeds. Chop half the melon into small pieces and blend to make a purée. Cut the remaining melon into small cubes or balls and set aside.

If using mango or soursop, make sure all seeds have been discarded. Purée half the fruit and cut the remainder into neat pieces.

Mix together the sago, coconut milk, puréed fruit, cubed or cut fruit and sugar syrup. Serve chilled.

Rujak

A savoury salad of sour or unripe fruits found all over Indonesia, Malaysia and Singapore, combined with a spicy, pungent sweet-sour sauce. Suggested fruits: half-ripe papaya, sour mango, water apple, Otaheite apple, cashew apple, salak and pineapple. Cucumber can also be added.

2 cups mixed sliced or diced fruit
1 teaspoon dried shrimp paste
1 tablespoon thick black shrimp paste (or 2 teaspoons dried shrimp paste)
1 heaped tablespoon tamarind pulp
1/4 cup warm water
2 tablespoons chopped palm sugar
2–3 fresh red chillies, chopped
1 tablespoon lime or lemon juice
1/2 teaspoon salt
1 tablespoon water

Put the fruit in a bowl. Wrap the dried shrimp paste in foil and toast under a grill or in a dry pan, turning after 2–3 minutes to cook on both sides until dry and crumbly.

Soak the tamarind pulp in water for 5 minutes, then squeeze and strain to obtain the juice.

Put both lots of shrimp paste, tamarind juice, palm sugar, chillies, lime or lemon juice, salt and water into a blender and process to make a paste. Pour over the fruits, toss and serve immediately.

Yam Taeng

This recipe from Thailand is a good way to use sour or unripe fruits such as half-ripe papaya, sour mango, water apple, cashew apple, Otaheite apple, salak, pomelo and pineapple. Fresh water chestnuts and cucumber can be added if liked.

2 cups mixed sliced or diced fruit
2 heaped tablespoons cooked prawns or pork, finely sliced
2–3 shallots, sliced and fried until crisp and golden
2–3 cloves garlic, sliced and fried until crisp and golden
fresh coriander leaves to garnish

Sauce:
2 tablespoons lemon juice
2 teaspoons sugar
1 tablespoon Thai or Vietnamese fish sauce
$^1/_4$–$^1/_2$ teaspoon chilli powder or chilli flakes

Put the fruits and prawns or pork in a bowl. Mix together the **sauce** ingredients, stirring until the sugar dissolves. Immediately before serving, toss the fruit with the sauce, garnish with fried shallots, garlic and coriander and serve.

Index by Scientific Name

Index by Common Name